U0024974

拼布詩

詩與拼貼藝術的對話

張蘭芳

推薦序 *foreword*

縱谷稻浪、湛藍海岸形成花東豐富的自然環境和人文歷史，孕育出美麗的文化風貌，是發展藝術文化，創造在地美學的寶貴元素與資產。張蘭芳女士為臺東在地資深藝術家，創作生涯近50年，擅長以碎布角料拼貼作畫，採金針、鳳梨、稻浪等主題，透過樸拙的雙手，展現東部田園風光及大地景色，近年多將短詩與貼布畫結合，創作屬於臺灣的拼布詩，主題多以植物、動物及昆蟲為創作理念，搭配短詩，達到療癒、趣味的效果。

本作品構圖看似簡單，但用色直接且貼切，透過書籍，我們感受到藝術家雖已高齡85歲，但仍以質樸的素材展現對於生活的觀察與童心，令人敬佩，推薦大家仔細欣賞。

The rice waves of East Rift Valley and the azure blue coastline form the various natural environment, the humane history and beautiful culture of East of Taiwan. This is the precious essence and asset of developing art and culture and of creating local art. As an experienced local artist, Lan-fang Chang has been dedicated to art creating for almost 50 years. She's talented at making collages with rags, and through her simple hands she demonstrates the life of the countryside and the landscapes of Taitung, such as plucking daylilies and pineapples and rice waves etc. In rent years, she's combined the short poems with collages and made them the collage poems that belong to Taiwan. The themes of her works include plants, animals, and insects. Being collocated with short poems, it achieves healing and fun.

The compositions in the book look uncomplicated, but the colors are direct and appropriate. Though the artis is 85 years old, through this book we can feel the observation of life and a young heart of the artist be demonstrated with simple materials. It's admirable. I recommend appreciating it carefully.

國立臺東生活美學館館長　江愚
National Taitung Living Art Center　JIANG,YU

Lang-fang was born in Changhua and migrated to Taitung with family at the age of 8. The beautiful nature and the simple people of Taitung are what I always want to record. Communicative people are suitable for writing prose, and I, as a taciturn person, just want to write some short poems to express the memory for Taitung in my heart.

Japanese poet Shibata Toyo started to write poems at the age of 92 and published her first collection of poems at the age of 98. It seems to remind me that I am still young, and I should write more. I am thankful that this collection of poems can be published due to the encouragement of my niece Nai-wen, the translation from my niece Chin-ya, the recommendation of the CEO Zong-kai Hung, the assistance from National Taitung Living Arts Center, and the help of my daughter Wei-Hsiu. Hope you like it.

蘭芳出生於彰化市，八歲隨著家人移居臺東，臺東有美麗的大自然和樸實的人民，一直是我想記錄下來的。健談的人適合寫散文、小說，而我，木訥寡言，只想寫幾句短詩來表達我心中對臺東記憶。

日本女詩人柴田豐92歲寫詩，98歲出版自己的第一本詩集，似乎在提醒我，加油呀～你還年輕呢，趕快再寫吧。能出版這本詩集，要感謝姪女乃文的鼓勵推薦、姪女琴雅的協助翻譯，洪宗楷執行長引薦，還有臺東生活美學館的協助，最後是女兒為琇統籌，才能讓我完成出書的夢想，希望你們會喜歡。

作者簡歷 *Author Bio*

布作家

張蘭芳

Chang, Lan-fang

85歲素人藝術家張蘭芳透過樸拙的雙手，以畢生熱情與執著創造另類的環保藝術，運用碎布角料拼貼作畫，創作題材來自臺東大地的自然元素。農作物採收、四季更軼、動植物細微的變化，一一記錄下東台灣的美好印記。

蘭芳參展無數，2007年受邀參加臺東美術館開幕首展，近期醉心於日本俳句體創作中文短歌，佐以布貼插畫，充滿童趣的文字與色塊遊戲。作品集樂活療癒、永續綠能、友善環境為創作訴求，破碎與重組的巧思，拼貼出臺東農村之美。

85-year-old amateur creates a new type of art works which are about eco conservation by her very hands with her whole heart and passion. She collages rags, and the themes of her works come from the natural materials from the land of Taitung. She has recorded the pretty things in her mind which include harvests, the changes of seasons and the variations of animals and plants.

Lang-fang has attended many exhibitions. In 2007, she was invited to attend the opening exhibition of Taitung Art Museum. In recent, she has been obsessed with writing Chinese poems which are in the form of Japanese haiku, and illustrating with collages makes it an interesting game of words and colors. The works are about healing and LOHAS.

蘭芳工作室

Lang Fang Studio

空巢靜靜的
大自然的另一方
白雲處處飄

Being silent in the empty nest,
On the other side of nature,
White clouds floating everywhere.

Near the pond in a rainy night,
Ribbit, ribbit, ribbit, ribbit, ribbit, ribbit,
Frogs reciting poems happily.

夜雨池塘邊
咯咯咯咯咯咯
青蛙吟詩樂

土裡閉關修
出土歌頌經典勤
禪師你好

Retreating in the soil and
Singing classic songs diligently after coming out from the soil,
Let’s greet the master cicada.

秋風吹又吹
白色蘆葦搖啊搖
雲雀空中叫

Autumn wind blowing,
White reeds waving,
Skylarks tweeting in the sky.

心形高高掛
百貨公司屋頂上
母親節將到

Heart-shaped decoration hung up high
On the roof of the department store–
Mother's Day.

大地池塘中
蜻蜓點水漣漪
盛夏的正午

In the big pond of the land,
Dragonflies skimming the water
surface to make ripples
At noon in midsummer.

飛來又飛去
花叢中嗡嗡嗡嗡
蜜蜂採蜜忙

Flying back forth,
Buzzing in the blossoms,
Bees are busy gathering honey.

深夜靜靜的
貓頭鷹咳咳嗽
老鼠嚇跑了
Quiet late at night,
An owl coughing,
A rat frightened to escape.

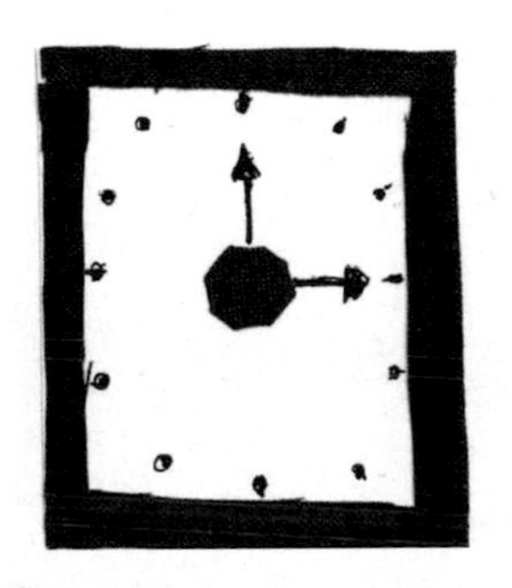

遵守時間
太陽有道時針相隨
永不怠忽

Always being punctual, the sun,
With a hand of a clock by its side,
It never neglects.

喜舞不善歌
一身是花花綠綠
蝴蝶妳是誰

Liking to dance and being not good at singing,
Always wearing a colorful dress,
Who are you, butterfly?

咕咕　咕咕
弟弟說不是我
鸚鵡笑了

Coo coo, coo coo,
"It's not me." Little brother said,
The parrot laughed.

長長一條線
風箏飄飄的走了
傷心又斷腸

A long long string,
The kite flew away,
Feeling sad and heartbroken.

矗立淨心
鳥大便於頭上也無語
偉人銅像

Established to purify hearts.
Being silent when a bird defecates on its head –
A statue of a somebody.

紅蛋紅紅的
一個一個盛一盤
嬰兒滿月了

Putting the red eggs
In the plate one by one–
The newborn baby’s first month.

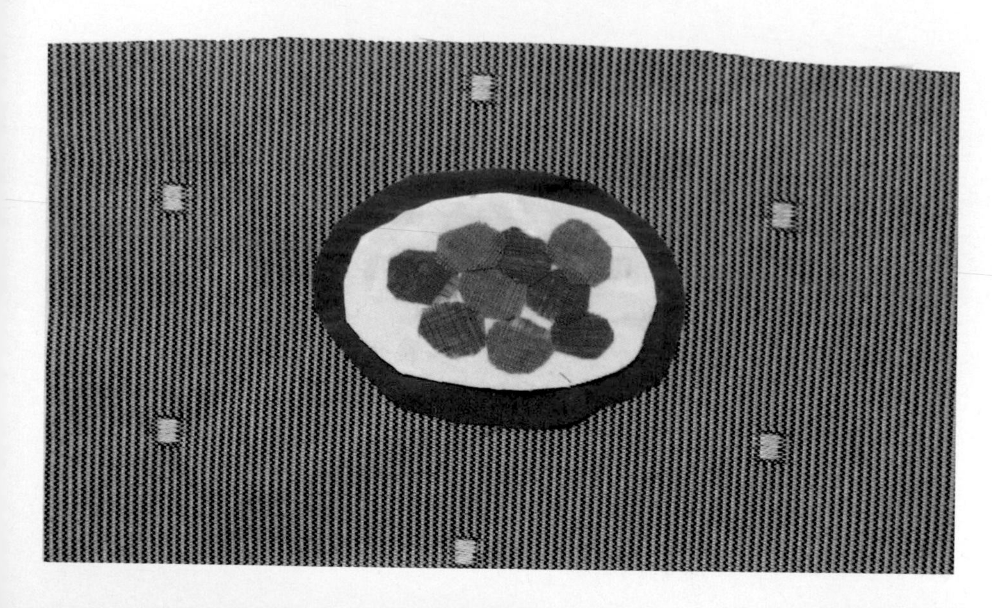

自由，自由，

氣球高空飄遠，

走向彼岸？

Freedom, freedom!

Balloons fly high away

To the other shore?

美麗庭園
自由奔放的舞者
藤蔓植物

In a beautiful garden,
Free-spirited dancers–
Vines.

哆哆哆哆哆
哆出一洞又一洞
啄木鳥的家

Do do do do do,
Pecking holes one by one–
The houses of woodpeckers.

門前麻雀叫
電線桿上五線譜
嘰嘰喳喳喳

Sparrows tweeting in front of the door
Music staffs on the telephone poles
Chirp, chirp, chirp, chirp, chirp.

農民豐收季，大群麻雀來道賀，曬穀場熱鬧非凡

In the season of harvest
Flocks of sparrows come to congratulate,
It’s hustle in grain shovels.

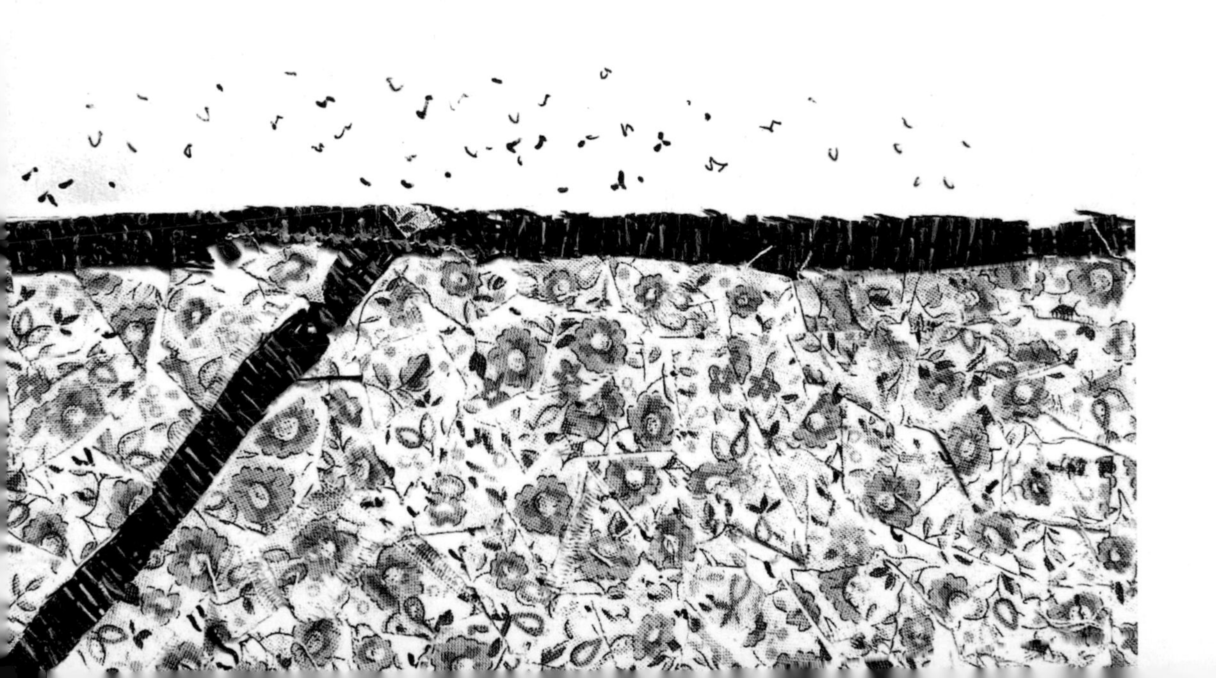

呱呱呱呱呱
洞口傳出來的聲音
貓豎耳恭聽

Ribbit, ribbit, ribbit, ribbit, ribbit,
The sounds coming from the hole
The cat listening carefully.

向日葵仰望，在美好的光彩中
遍地是榮耀

Sunflowers looking up in brilliance,
It’s glory everywhere.

很熱的夏天
帥哥蜻蜓戴墨鏡
老大你來了

Hot summer
Handsome dragonfly wearing sunglasses –
Here comes the boss.

糟糕桌子上
螞蟻宴客坐滿滿
糖罐破一洞

Oh! No.
Lots of ants gathering on the table –
A hole on the sugar jar.

默默揮別
一生一世的交代
竹子開花

Saying farewell silently
An explanation of all its life –
Bamboos blossom.

屋簷下
鴿子咕咕響不停
談情說愛

Under eaves
Pigeons never stop cooing –
Flirting with each other.

屋簷一角
天亮叫人起床
小燕四隻

At the corner of the eave
Waking up people at dawn –
Four little swallows.

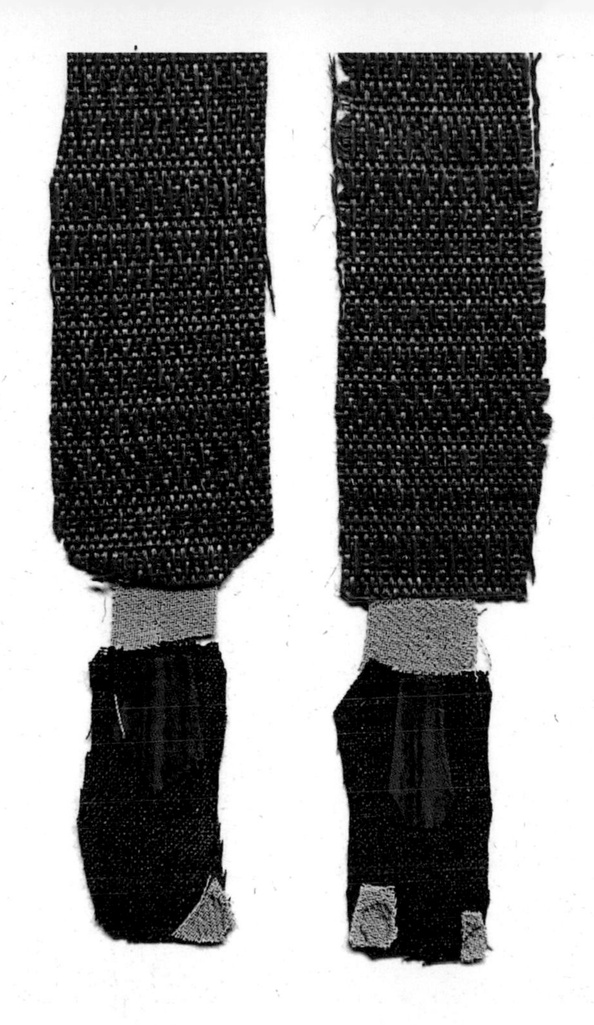

走著走著
五指嚮往外面風景
鞋破個洞

Walking for a while
Five toes yearn for the scenery outside –
The shoes worn out.

自然界一方
仙人掌寂植荒野
點綴綠意

An area of the natural world
Cactuses grow in the wilderness silently
Adding some green.

月娘含羞
時露臉時遮面
悄然夜行
Shy moon
Sometimes showing up,
Sometimes covering its face,
Moves silently.

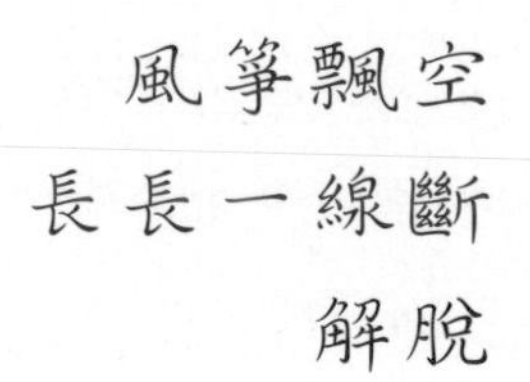

風箏飄空
長長一線斷
解脫

A kite flying up to the sky
The long line being broken–
Relief.

鳳凰花燦爛
毛毛蟲來騷擾
她傷心落淚

A poinciana blooms,
Caterpillars come to harass
That makes her sad and tear.

商街冷清清
唯獨鞋店老闆笑嘻嘻
蜈蚣上門來

Not many customers on the shopping street,
Only the shopkeeper of the shoe shop is
smiling broadly–
A centipede is coming to the store.

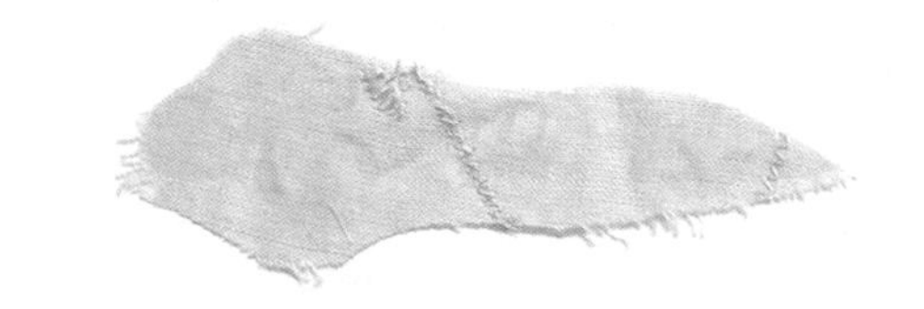

海水退潮時
鵝卵石高興啦啦響
可以下水了

When tide goes out,
Cobblestones are happy clacking.
It's time to go into the water!

生居松樹林　死後長眠琥珀中
小蟲很值錢

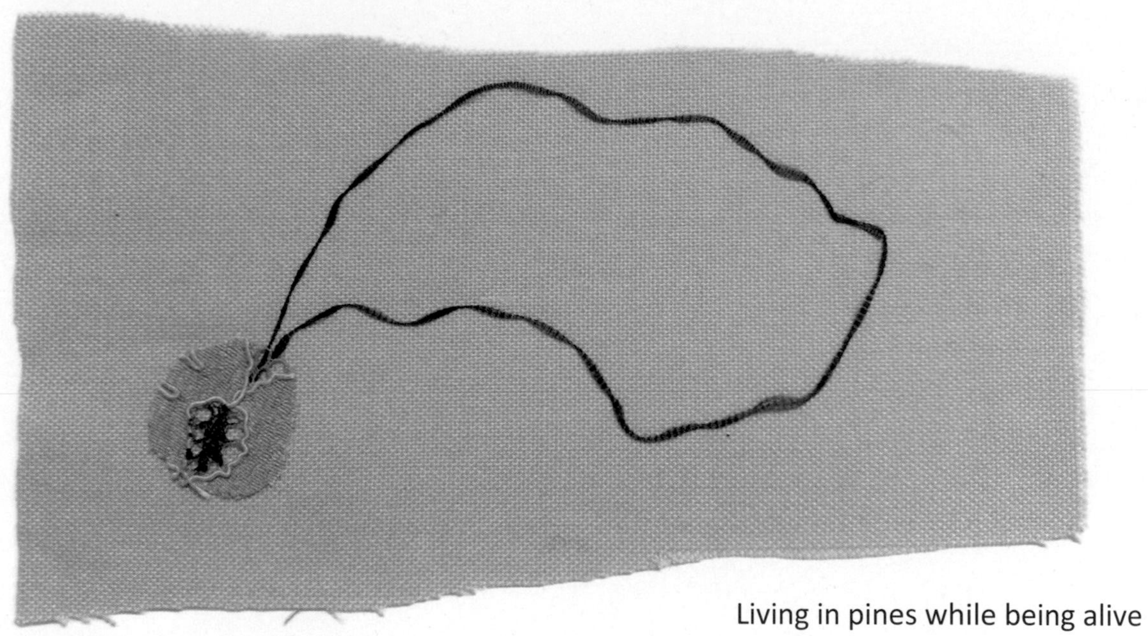

Living in pines while being alive
Sleeping in amber forever while being dead,
Valuable are the little bugs.

舊鐵道旁，波斯菊盛開並說，歡迎光臨

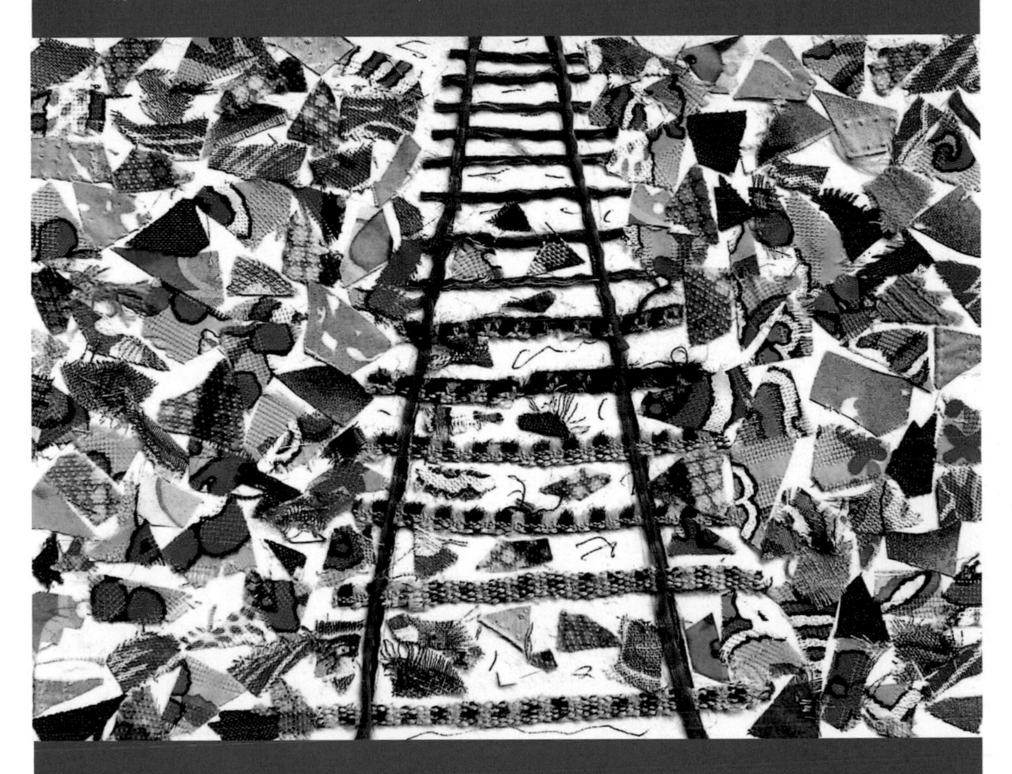

Beside the old railway
Cosmos bloom and say,
“Welcome here!”

必自立更生
因價值還沒被發現
我是草

Having to be self-reliant because
My value hasn't been found yet,
I am grass.

遠望西邊
傍晚紅色彩霞
群鳥歸巢

Looking over to the west
Sunset clouds at nightfall,
Flocks are flying back to nests.

漁夫歡笑
魚瞪眼死不瞑目
漁港一景

Fishermen laughing happily
Fish cannot die in peace –
A scene of a fishing port.

憐香惜玉，清高而不傲，含羞草

Kind,
Self-contained, but not arrogant –
Mimosas.

路燈亮亮
守宮飛簷走壁
飛蛾死翹翹

Bright are streetlamps
Geckos leaping onto roofs and vaulting over wall,
Moths die.

桌上佳餚，紅燒魚白眼帶恨，很難下嚥

Tasty dishes on the table
Braised fish's eyes with hatred
Hard to swallow it

大樹落葉了
小草感恩葉當被
禦寒好過冬

Leaves falling down from big trees,
Grass takes it as a cover with gratitude
To keep warm during winter.

早安我來了
再見我走了
木屐咯咯聲

"Good morning." I am coming.
"Good-bye." I am leaving.
Getas' sound.

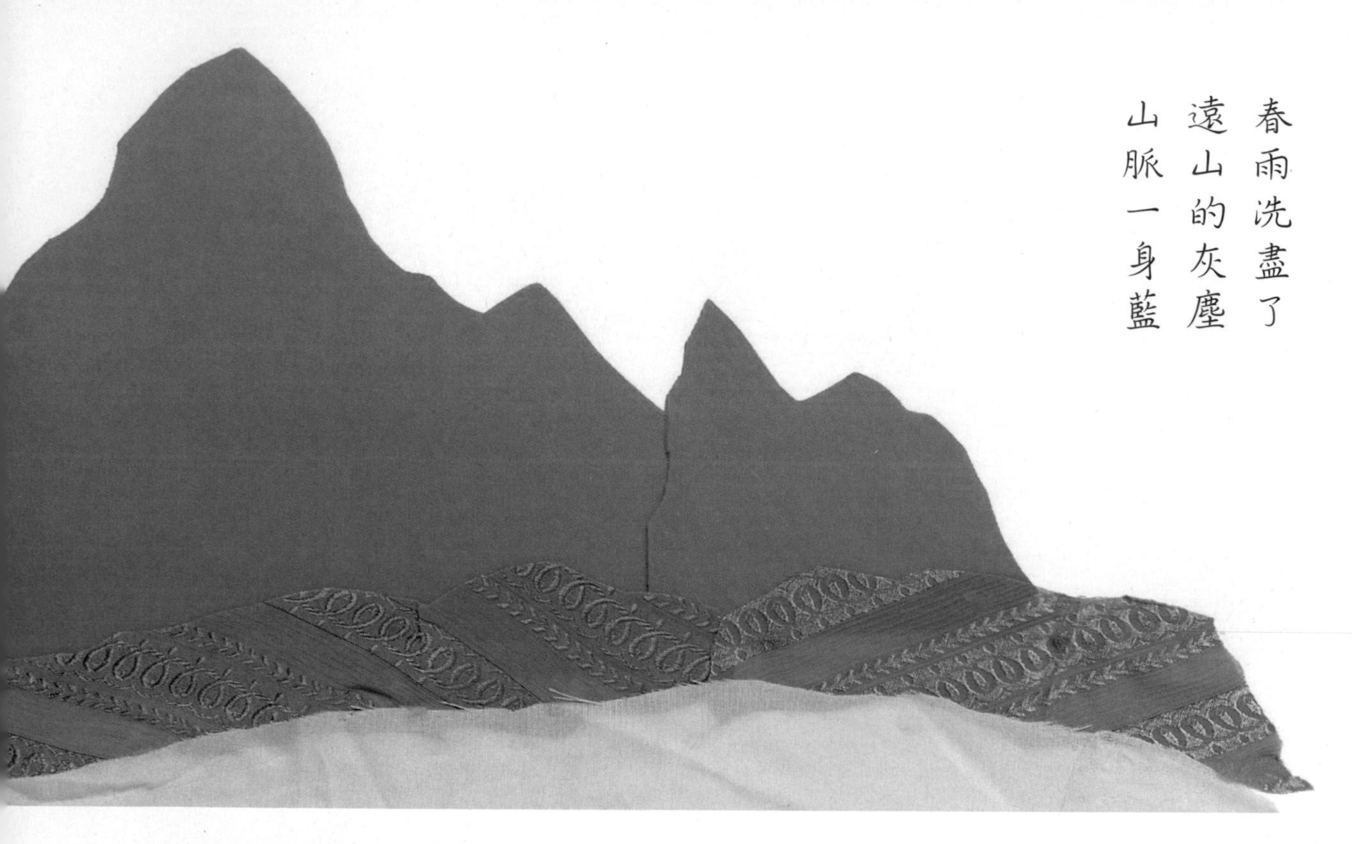

春雨洗盡了
遠山的灰塵
山脈一身藍

Spring rain washes away
The dust around the faraway mountains
The mountains all in blue

春天梅雨季
樹下朵朵香菇傘
美麗又可愛

Rainy season
Umbrella-liked mushrooms under trees
Pretty and adorable

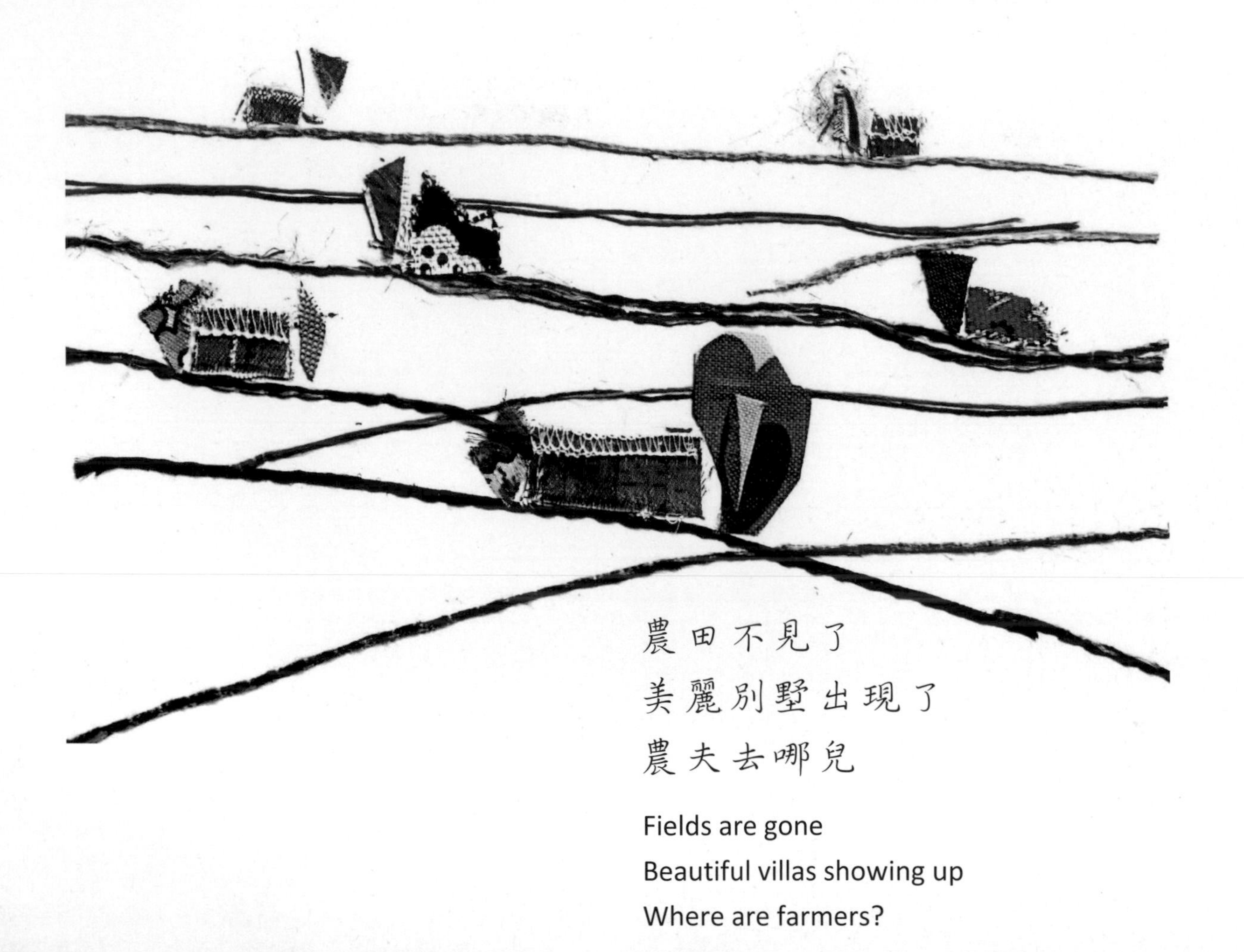

農田不見了
美麗別墅出現了
農夫去哪兒

Fields are gone
Beautiful villas showing up
Where are farmers?

太陽日日巡
向日葵歡喜相隨
和風徐徐吹

The sun patrols every day,
Sunflowers like to accompany it.
Breeze blows.

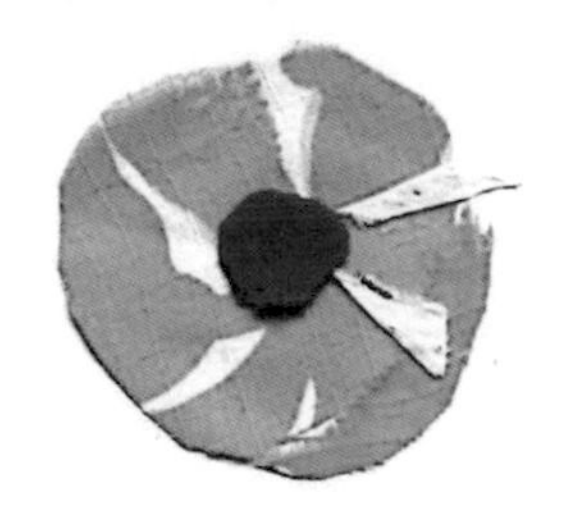

原形雖沒變
氣息已經不存在
書中的壓花

The original form never changes,
The flavor is gone –
Pressed flowers in books.

媽媽的手
年輕時胸前忙
年老時腰後閒

Mother’s hands
Busy while being young,
Free while being old.

我犯何罪呢？

採、曬、掰、揉、炒、押、泡，

不錯是好茶

Why should I be punished?

Picked, basked, broken, rubbed, fried, pressed, brewed,

Nice tea!

秋菊盛開
母親喜歡的花
很想念

Chrysanthemum blooms
Mother’s favorite flower.
Miss her so much.

春天到　都蘭山穿新娘裝
白雲處處飄

Here comes spring,
Mt. Doulan wearing a wedding gown,
White clouds floating everywhere.

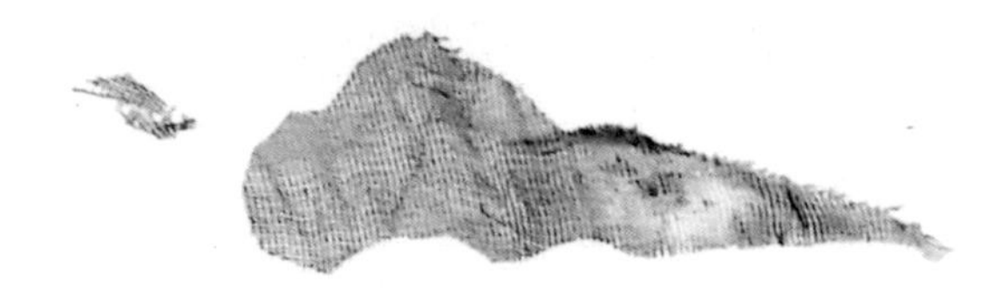

晴空萬里
地上煙小妹想找
雲朵作朋友

Cloudless sky,
Little smoke wants to find
Some clouds to make friends.

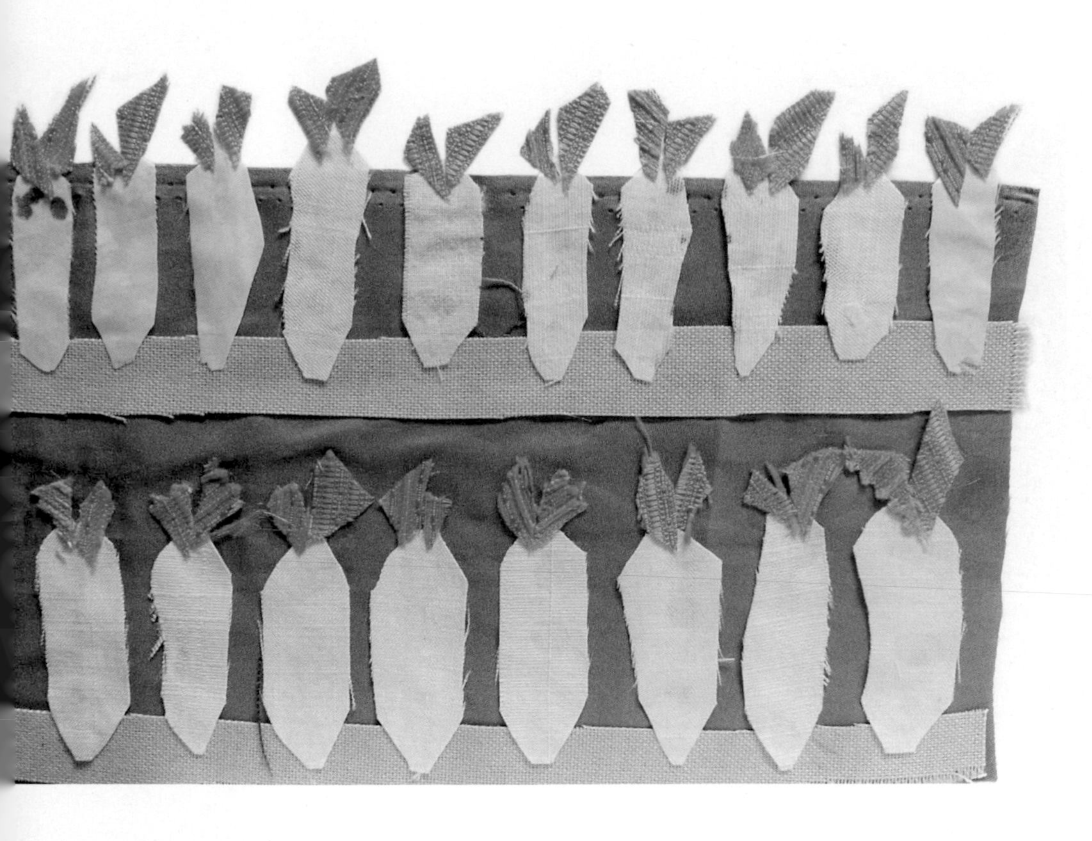

白色小蘿蔔
可愛站滿菜架上
像船上海軍

Cute little white radishes
Standing on the rack
Like navy soldiers

被寵愛的貓
是等待吃的時間
過活也

A loved cat
Always waits for eating
Just for living

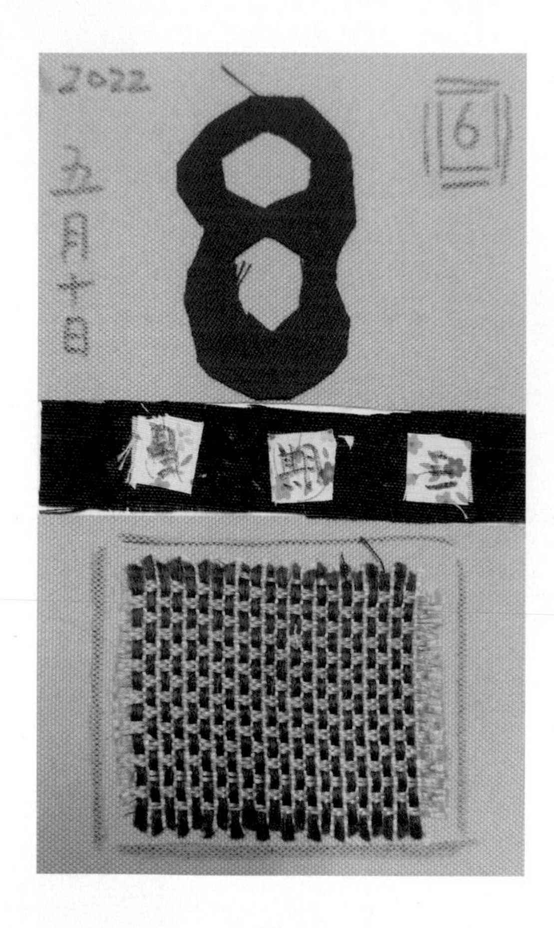

一年出場一次
個個有機會
日曆

Showing up once a year
Everyone has a chance –
Daily calendar.

甜蜜關係
杯子與嘴唇相吻
感覺在心頭

Sweet relationship
The cup and lips kiss
The feeling is inside hearts

熟悉的音樂
穿梭大街小巷中
垃圾車來了

Familiar music
Heard all over the city
The garbage truck is coming

入鄉隨俗
日本溫泉
赤裸泡湯

Do as the Romans do,
In Japanese hot spring
Bath nakedly.

拼布詩

詩與拼貼藝術的對話

作　　者　張蘭芳
譯　　者　黃琴雅
企劃編輯　莊為琇
美術編輯　卓文怡

出版單位　蘭芳工作室
臺東市漢陽北路169巷2號（展售處）
089-335365
協辦單位　國立臺東生活美學館

印　　刷　良品工作室
初版一刷　2022年7月
定　　價　新台幣350元

Poems of Collage

The Conversation Between Poems and the Art of Collage

author Chang, Lan-fang
translator Huang, Chin-ya
Project editor Chuang, Wei-Hsiu

Contact for purchase **Lang-fang Studio**
Address No. 2, Ln. 169, Hanyang N. Rd.,
Taitung City, Taitung County 950015,
Taiwan (R.O.C.)
Tel 089-335365

國家圖書館出版品預行編目（CIP）資料

拼布詩 詩與拼貼藝術的對話＝Poems of collage : the conversation between poems and the art of collage/ 張蘭芳作， --初版-- 臺東市：蘭芳工作室：2022.7，
64面；21×14.8公分；中英對照；
ISBN 978-626-96365-0-1（精裝）

1.拼布藝術 2.短詩 3.作品集

863.51　　111011192